AEI Special Analyses
Timely examinations of issues being debated in the formulation of public policy

The Problem of Military Readiness

Melvin R. Laird
with Lawrence J. Korb

American Enterprise Institute
for Public Policy Research
Washington, D.C.

Melvin R. Laird is chairman of the AEI Public Policy Project on National Defense. He served as secretary of defense from 1969 to 1973 and was a member of the House of Representatives Appropriations Committee, Subcommittee on Defense, from 1953 to 1969.

Lawrence J. Korb is resident director of defense policy studies at AEI and co-editor of the *AEI Foreign Policy and Defense Review*.

ISBN 0-8447-1087-3

Special Analysis No. 80-3

Printed in the United States of America

Contents

Preface

Our national defense establishment has many critical problems. Earlier this year, the American Enterprise Institute for Public Policy Research published *People, Not Hardware,* a report on the most important problem of national defense—the failure to attract and retain adequate numbers of qualified men and women for military service. The second most important problem, the inadequate readiness of our armed forces, is the subject of this report. A third report, currently in preparation, will outline the requirements of the defense budget for fiscal year (FY) 1982 in order to meet personnel, readiness, and hardware needs through 1990. We expect this report to be released in the autumn.

1

Introduction

This nation raises and maintains military forces for two primary purposes: to deter others from employing military force against us and to conduct military operations if deterrence should fail. The ability to be ready to fight or conduct military operations successfully is by far the more important of the two functions. If our military forces are not prepared to fight, then few of our adversaries will be deterred from committing hostile acts against us or our interests. As we enter what many observers call the dangerous decade of the 1980s, the international situation is in profound disarray and the possibility that this nation will have to employ military force grows ever more likely as we are confronted by more and more situations adverse to our interests. The Persian Gulf area, upon which we and our major allies depend for the vast majority of our imported oil, has not yet recovered from the downfall of the shah. Sub-Sahara Africa, upon which we depend for such strategic minerals as bauxite, chromium, and manganese, is torn apart by racial turmoil and the presence of Cuban and East German proxies. The Soviets, flushed with their success in Cuba and Afghanistan, may grow more adventuresome and could try to test us in other more critical areas. Central America, in our own backyard, is exploding in revolutionary turmoil that could threaten U.S. security.

Yet at this precise moment there is understandable and growing concern about the capability or readiness of our forces to conduct military operations. Almost daily, our newspapers and newscasts are filled with stories about the inability of ships to get under way, aircraft to get off the ground, and tanks to move. In the past few months, the major networks have run several specials about the current lack of preparedness of our armed forces. The television journalists, through interviews with enlisted men and officers, congressmen and civil servants, Republicans and Democrats, convey the impression that our military forces cannot adequately perform their primary functions. The inability to place six helicopters into Tehran seemed to confirm this impression. Instead of having a well-oiled military machine befitting a superpower, we appear too often to have "the gang that can't shoot straight."

Not everyone agrees that our military forces are as unprepared as this might indicate. Many people talk confidently about the success of the all-volunteer military, the readiness of U.S. forces to secure access to vital resources, and the capability of our forces to meet the Soviet threat. Secretary of Defense Harold Brown, for example, noted on July 28, 1980, in a speech to the Commonwealth Club of San Francisco, that those who suggest that U.S. forces are unprepared are "playing fast and loose with the truth."

Is it true that if necessary we are not at this time ready to go to war? Can it be possible that a nation that now spends more than $150 billion per year for defense does not have the capability to apply military force successfully in pursuit of its national objectives? How, short of engaging in war, can we determine whether our forces are unprepared? How do we define readiness or military capability? Are they the same? If not, how do they differ? Is it important to differentiate them? Are the critics and supporters of our military forces focusing on strategic or nuclear forces as well as general purpose or conventional forces? Or are they concerned only with the latter? If so, are all of our conventional forces in the same condition?

Part of the problem in analyzing the current state of military preparedness is a semantic one; that is, it is attributable to a confusion of terms. Some people who discuss the issue talk about the capabilities of our armed forces to achieve certain objectives. Others talk about the readiness of our forces to undertake certain missions. Still others use capability and readiness interchangeably. In fact, capability and readiness are not the same. Capability is the more inclusive concept. Readiness, on the other hand, is a component of capability. But both are important.

To assess adequately the capability of all or part of our armed forces to carry out certain functions, one must analyze four separate components:

- *Force Structure:* The number and type of major units currently possessed by the armed forces; for example, the number of ground divisions in the army and Marine Corps, the number of tactical air wings in the air force, or the number of ships in the navy.

- *Modernization:* The rate at which the nation is replacing or adding to its major equipment; for example, the number of tanks, planes, and ships currently being procured in the annual defense budget.

- *Sustainability:* The ability of the force structure to conduct military operations long enough and with sufficient intensity to achieve its objectives. Sustainability focuses on such areas as the amount of ammunition and spare parts currently possessed by the armed forces, the ability of the nation to keep its deployed forces adequately supplied, and the mobilization base of the country.

- *Readiness:* The ability of the currently configured force structure to perform its assigned missions promptly. Readiness is concerned with such issues as the ability of a tactical air squadron to deliver bombs to a target or to engage in anti-aircraft warfare, or the ability of a destroyer to conduct anti-submarine warfare.

Breaking down military capability into its four components clears up part of the confusion about the present condition of our armed forces. As these definitions indicate, the armed forces can be ready but not capable for a variety of reasons: for example, because the force structure is too small, the amount of ammunition on hand is inadequate, or the pace of modernization is too slow. Thus, even if all of the 400 general purpose ships in the navy were ready to perform their specific functions, the capability of the navy to carry out its mission of sea control in three oceans would be in doubt because it would not have what many believe is the requisite number of ships, that is, 600. Similarly, even if the navy possessed a force structure of 600 ships, it might not be able to carry out its mission if too many ships lacked trained personnel or properly functioning equipment.

Confusion also results from mixing strategic and general purpose forces together in blanket appraisals. For example, while the tactical aircraft in a fighter wing, which carry conventional weapons, may lack spare parts, the same might not be true of the B-52s, which are configured to carry strategic nuclear bombs and cruise missiles. Similarly, a shortage of trained personnel on a navy oiler does not necessarily mean that the Poseidon submarines, which carry sea-launched ballistic missiles, also are short of critical people.

This study will analyze each of the four components of capability as it applies to the conventional or general purpose forces of the United States in order to ascertain whether these forces can carry out their objectives. If they cannot, we will consider why not and what can be done about it. In our view, there is no more critical issue facing the United States as we enter what will surely be a challenging decade for the West.

Although we will not examine in this study the capabilities of our strategic and theater nuclear forces, it is important to keep in mind that, unlike the situation that prevailed until a decade ago, we have some serious deficiencies in each of those areas. In the strategic area, we find that our intercontinental ballistic missiles (ICBMs) are becoming vulnerable, our submarine-launched ballistic missile (SLBM) production is lagging, our bombers are aging, and our strategic balance with the Soviets is being upset. By the middle of this decade, the Soviets will have an advantage in every single static indicator of the strategic situation.[1]

Similarly, in regard to theater nuclear weapons in Europe, our balance with the Soviets currently is precarious. Although this does not mean we would lose a nuclear war, it does mean that our conventional capabilities must be adequate. If they are not, we can only achieve our objectives militarily by escalating to nuclear warfare. Moreover, even if we did have a strategic or nuclear balance, such forces will not be relevant when it comes to the critical area of protecting the oil supply, upon which our economic existence depends, from internal disruption or from conventionally armed Soviet proxies.

2

Force Structure

The active-duty general purpose forces of the United States may be grouped into four major categories: ground forces, naval forces, tactical air forces, and mobility forces. The army and marines provide the ground forces, ships come from the navy, tactical air power is provided by the air force, navy, and Marine Corps, while airlift is the province of the air force, and sealift the responsibility of the navy and Maritime Administration. At present the United States has nineteen ground divisions (sixteen army and three marine divisions), thirty-seven airlift squadrons, and 200 sealift ships in the Military Sealift Command or in the National Defense Reserve Fleet.

Assessing the adequacy of these numbers requires that we understand their mission and the threat that these forces are likely to encounter in trying to accomplish their objectives. On the basis of these two criteria, it is clear that the U.S. force structure is inadequate. Thus, there is a question about its ability to carry out its objectives. Looking at each of these force categories individually makes it clear that our current force structure is simply too small.

The sixteen army divisions are configured primarily to fight on the Central Front in Europe. Five of the army divisions are stationed there and another five currently in the United States are supposed to arrive in Europe within ten days after mobilization. The Soviet Union alone has no less than fifty combat-ready division-equivalents[2] that it is prepared to throw into the land battle in Europe. While it is true that U.S. forces will be assisted by the ground forces of our NATO allies, the Soviet Union can also call on its Warsaw Pact satellites. The combined ground forces of the Warsaw Pact in the NATO guidelines area total 935,000 while those of NATO number only 732,000, a difference of over 200,000. There is little doubt that the army should have more active forces if the Western alliance is to have a reasonable assurance of handling a conflict in Europe.

Not only are the ground forces of the NATO nations outnumbered, they are also outgunned. The Warsaw Pact has a 3-to-1 advantage in the number of tanks, a 4-to-1 advantage in the number of artillery pieces, and a 3-to-1 advantage in the number of tactical missiles. The U.S. army, by its own admission,[3] has only 78 percent of the tanks that its undersized force structure is supposed to have and is short approximately 60,000 wheeled vehicles.

One of the army's six remaining divisions is deployed to Korea. Two others are on the West Coast and can be used to reinforce Korea or can be sealifted to Europe to arrive about a month after mobilization. The remaining three divisions are allocated to the newly created Rapid Deployment Force (RDF). If a future conflict can be confined to a single theater, the army's present non-European active force structure may be adequate. But if fighting occurs simultaneously in more than one theater, these six divisions would be hard pressed to handle their assigned tasks.

The three marine ground divisions have enough missions to keep at least ten divisions busy. One division is earmarked for the RDF, which would most likely be used in the Middle East or Persian Gulf area. The two other divisions are allocated to the European theater for defense of the Northern Flank. The supreme allied commander in Europe would also like two marine divisions to throw into battle on the Central Front as a reserve force. The commander-in-chief in the Pacific needs at least one marine division for use in an amphibious role in a Pacific campaign or to balance the 300,000 Soviet ground troops stationed in the Far East. Finally, the Atlantic commander would like to be able to count on the marines for possible use in Central and South America or in sub-Sahara Africa.

Our naval forces are in much worse shape numerically than our ground forces. The 400 general purpose ships are at best a one-and-one-half ocean navy; that is, they can maintain an up-to-strength force forward deployed in one ocean and a small force in another. At present the navy is being asked to undertake a three-ocean commitment, that is, to maintain a viable forward presence in the Mediterranean, the Indian Ocean, and the Western Pacific. To do this adequately, it needs at least an increase of 200 ships, bringing the total to 600.[4] Because it has only 400 ships, the navy has had to cut in half its carrier presence (and their escorts) in the Mediterranean and in the Western Pacific and to withdraw its only amphibious force from the Mediterranean, thus leaving those areas considerably short of the needed presence and firepower. The navy is currently stretched so thin that it has absolutely no margin for error. For example, when the carrier *Midway* was disabled by an accident in the South China Sea on July 29, 1980, no operational carriers were left in the Western Pacific. There is no doubt that the navy cannot continue to meet its peacetime commitments without a dramatic increase in size.[5] Moreover, if a war with the Soviet Union should erupt, the navy would have to withdraw its forces west of Hawaii and swing them into the Atlantic just to carry out its missions in that theater, thus abandoning the Western Pacific to the 200 ships in the Soviet Pacific Fleet and precluding any hope of opening a second front to relieve the pressure on Central Europe.

The air force presently has twenty-four tactical air wings. This is two short of its minimum goal of twenty-six tactical air wings and nine short of the thirty-five wings that it needs to have a "reasonable assurance"[6] of carrying out its mission. Like the army, the air force is oriented primarily toward a European contingency where its job will be the interdiction of Soviet forces and the support of our army ground forces. Twenty-nine of its seventy-seven fighter squadrons, or 38 percent, are actually in Europe, while another thirty-eight squadrons, or 50 percent, are earmarked to be moved from the United States as soon as hostilities erupt. Only ten fighter squadrons are in the Pacific. However, even with heavy European emphasis, our air force together with our allies would still find itself outnumbered in Europe by more than 2-to-1 by the Warsaw Pact, 7,500 aircraft to 3,300. Moreover, the air force has also been tasked to provide up to four tactical air wings to the Rapid Deployment Force. Since these would come from those U.S.-based units targeted for Europe, they would not be available for use by the RDF if a war occurred in Europe and the Persian Gulf simultaneously.

The navy's twelve air wings are only enough to place one on each carrier. Ideally, in order to provide for combat losses, each carrier should have an additional one-third of an air wing assigned to it, or a sixteen-carrier air wing force. Similarly, the Marine Corps' three air wings allow only one to support each marine ground division. As noted above, however, there are demands for far more than the three marine ground divisions. And since each division is supposed to have its own air power, there would also be a need for more air wings.

While our ground, sea, and air forces all have numerical shortfalls, none is in as bad a shape as our mobility forces, particularly our strategic airlift forces. The air force has 280 C-141 and 76 C-5 aircraft to carry men and materiel from the United States to the area of operations. These 356 aircraft must be capable of moving five army divisions and nine tactical fighter squadrons to Europe within ten days after mobilization. In addition, minimum essential support to both our ground and air forces must be provided by these aircraft until sealift can deliver more substantial support. This simply cannot be done by 350 planes in less than thirty days. While the use of the Civilian Reserve Air Fleet (CRAF) assets (34 wide-bodied cargo planes and 187 narrow-bodied passenger aircraft) would reduce the time needed for mobilization by about five days, the planning objective of ten days or less still could not be achieved. Nor is it certain that DOD could get access to all the planes in CRAF immediately.

Our airlift cannot even handle adequately the so-called minor contingency or half-war scenario. It would take the Military Airlift Command (MAC), using all its assets and augmented by CRAF, at least fifteen days to move the army's lightest divisions—the 82nd Airborne and 101st Airmobile—into the Persian Gulf. A mechanized brigade (8,000 men equipped with tanks and heavy artillery) would take more than twenty-two days, and a marine amphibious brigade (12,000 men with their equipment), about twenty-five days. A regular mechanized division of 17,000 people with their equipment could not be lifted to the gulf region in less than a month. Placing a significant force of about 200,000 men into the area would take six full months![7] Moreover, these time estimates assume that there are no demands on our mobility assets anywhere else in the world. If there are demands to move men and materiel to Europe or the Pacific simultaneously, the times noted above would increase dramatically.

Finally, the forces that can be moved into the Persian Gulf lack both the size and the firepower to handle most of the contingencies they are likely to face. There is no doubt that the RDF, even if it could get there in time, could not handle the Soviet Union in its own backyard.[8] But even the armies of the clients of the Soviet Union in that region have more tactical mobility and firepower than that now possessed by either the light army divisions or the marines. For example, the Iraqi army alone has 3,500 tanks and armored fighting vehicles, almost seven times the Marine Corps' entire inventory. Syria has about 2,500 tanks and even tiny South Yemen 260, five times more than an unmechanized U.S. Army division.

It is true that the Sealand ships that DOD has chartered this summer and dispatched to the Indian Ocean will reduce to a week the deployment time of some units to the Persian Gulf. These seven ships, however, can only support a brigade-size unit of 12,000 men, hardly a match for the forces that the Soviet Union or its clients can move into the area.

Sealift is in even poorer shape than airlift. DOD exercises demonstrate that we need 3,000 cargo ships to reinforce Europe properly. The navy has but thirty-seven vessels in its own sealift force, 75 percent less than a decade ago. The National Defense Reserve Fleet numbers but 160 ships. The U.S. Merchant Marine could not help very much. Over the past two decades the number of ships in our Merchant Marine has been cut in half. The United States, with 560 ships in its entire merchant fleet, now ranks tenth among the world's merchant fleets, and our portion of the world's gross tonnage is less than 5 percent. Moreover, our merchant fleet is old; more than one-third of our ships are more than twenty-five years of age.

The lift situation is compounded by the fact that our own navy lacks adequate lift for the marines. At the present time, the navy has only enough amphibious shipping to land one of the three marine divisions. To make matters worse, the amphibious ships are divided evenly between the Atlantic and the Pacific. Thus, it now takes six months to get ready to do a one-division assault in either area.

Examining major categories of the conventional arms balance between the United States and the Soviet Union demonstrates just how inadequate the size of the current force structure is. The U.S.S.R. leads the United States in total military manpower by 220 percent (4.4 million to 2.0 million), in tanks by nearly 500 percent (50,000 to 10,500), in artillery by over 200 percent (40,700 to 18,000), in tactical aircraft by 4 percent (4,350 to 4,164), and in major warships by 200 percent (523 to 260).

The inadequacy of our current force structure is also indicated by the way in which DOD has had to respond to the international ferment of the past year. When the Russian army brigade was discovered in Cuba in September 1979, President Carter responded to the new threat by creating the

Caribbean Contingency Joint Task Force (JTF). Although this is an imposing title, the task force consists of only seventy-eight staff officers who plan and conduct exercises in the Caribbean for the commander of the Atlantic Fleet. It has no combat forces assigned to it. Similarly, in response to the situation in the Persian Gulf and Southwest Asia, DOD created, on March 1, 1980, the RDF. Contrary to popular perception, the RDF does not consist of marines and army rangers ready to strike any place on the globe at a moment's notice. Rather, it is a planning agency with a staff of 250 people "separate but subordinate to" the Readiness Command. The RDF has no forces of its own. Thus, because its undersized force structure is already overcommitted, the nation has been forced to respond to the recent turmoil with symbolic rather than substantive actions. The JTF and RDF may give the appearance that something is being done, but in reality our force structure remains the same.

Obviously, these macrostatistics or indicators of force posture do not give the whole picture. One must also look carefully at the other indicators of military capability as well as the capabilities of our allies to determine who would prevail in a military confrontation between the superpowers. However, an inadequate force structure demands that there be offsetting advantages in the other three areas of modernization, sustainability, and readiness.

3

Modernization

There are comparatively few people who will contend that the force structure of our military is adequate to the task. But many of those who admit that our present force structure is inadequate believe that current and proposed programs will cure the inadequacies in our current posture in the not-too-distant future. For example, in January 1980 in his annual report to the Congress on our military posture, Secretary of Defense Brown noted that his proposed airlift programs "would enable us to make simultaneous deployments to Europe and elsewhere should the crises be concurrent (as is quite likely)."

While current defense programs will enhance our capabilities marginally in some areas, they will not result in a totally adequate force posture. Nor will they enable us to keep pace with the modernization efforts of our chief adversary, the Soviet Union. Moreover, in some areas, our force posture will actually deteriorate in absolute terms.

Over the next seven years, the number of tanks in the army's inventory will grow by 4,800, or just over 40 percent. While this growth is substantial, it will be less than one-third the planned increase in the size of the Soviet Union's tank force.[9] The army's growth in tank assets assumes that the trouble-plagued XM-1 tank overcomes its technical problems and reaches its maximum production rate of ninety per month in FY 1985 and is fully funded by the Congress. In FY 1981, DOD will be procuring fewer than 50 tanks per month. The Soviets now build over 200 per month.

Because of large authorizations for navy shipbuilding in 1975 and 1976, between now and 1985 the size of the general purpose U.S. Navy will increase by about 10 percent to about 450 ships. While this is hardly enough to maintain a three-ocean commitment, it will establish the high-water mark in fleet size for the remainder of this century. After 1985 the number of ships will decline rapidly to below 400 if the Carter administration continues its current rate of building 15 ships per year. To remain at its present size the navy needs to construct 17 new ships a year. To build and maintain a 600-ship navy requires an annual program of 22 ships.

Thus, the next decade will find the United States a seapower with less than a one-ocean navy. Moreover, during the next ten years, the navy of the landpower, the Soviet Union, will grow in size and capability because the Soviets are presently in the midst of an unprecedented qualitative improvement for their already sizable fleet. They are now building four classes of cruisers, all of which are superior to our own, and will add twenty-five new cruisers to their fleet over the next decade. Four of these will be 32,000-ton *Kirov*-class nuclear-powered cruisers. These will be the first nuclear-powered Soviet surface ships. In addition, the Soviets will add three aircraft carriers to their force during this decade. Two of these will be 40,000-ton *Kiev*-class carriers, while the other one is expected to be a 75,000-ton nuclear-powered aircraft carrier. Finally, the Russians are building a force of 13,000-ton *Ivan Rogov*-class amphibious ships, a new family of guided-missile destroyers, and twenty new frigates. Thus, by the end of the decade the Soviet navy will grow considerably in tonnage and firepower.[10]

Naval airpower will fare even worse than naval shipbuilding. In no single year over the past five years has the navy been allowed to procure sufficient aircraft to replace those lost to obsolescence or accidents, that is, 180 tactical aircraft annually. For FY 1981, the administration requested funds for only 104 planes, 76 planes or 42 percent short of the number needed to remain even. Thus, by the end of the decade naval aviation will be smaller or older or some combination of both. The most realistic estimate is that the naval air arm of the late 1980s will be 10 percent smaller than today and two-thirds of its planes will be over ten years of age. Moreover, even this smaller and older tactical air arm assumes that all 1,388 of the controversial F/A-18 Hornet aircraft will be purchased. Because the unit costs of this aircraft have risen by $8 million in the last year, the president is considering terminating the project. As Navy Secretary Edward Hidalgo noted in a recent memorandum to Secretary of Defense Brown: "The escalating costs of the F/A-18 have raised serious questions in my mind."

Marine Corps aviation is also in serious straits. It will drop in size by about 15 percent between now and 1987, even if it can afford to purchase its allotted portion of the F/A-18, which is now running at well over $20 million per plane.

Compared with naval and Marine Corps tactical aviation, air force tactical air power is in relatively good shape in the short term. Between now and 1986, it will grow by 35 percent and increase from twenty-four to twenty-six wings. Moreover, by the middle of the decade, only one-third of the fighters in its inventory will be more than ten years old. However, unless a new generation of fighters is developed now, the air force will experience in the 1990s the same problems that the navy and Marine Corps are currently facing. The size of the tactical air force will actually begin to decline after 1986 as the F-15, F-16, and A-10 production lines are closed out. Finally, the near-term growth in air force assets, substantial as they are, will not offset the planned growth in the size of Soviet tactical air forces. The Soviets build 1,150 fighters annually, compared with 500 for the United States, and will begin production of three new fighter aircraft within the next five years.

The size of the Marine Corps will remain the same, but the ability of the navy to lift or transport it will decline rapidly. By the middle of the next decade, the number of amphibious ships will drop to twenty-six, a decline of thirty-seven, or 59 percent, from present levels, which are already only half of what is needed. The marines of the mid-1990s may have to go ashore in rented rowboats if they wish to conduct an amphibious operation.

The one area that is expected to grow substantially is our strategic lift forces. The 280 C-141s will increase their capabilities by 30 percent, that is, from being able to move 2,170 tons per day to 2,800 tons, through a program of stretching each aircraft by twenty-three feet. In addition, DOD hopes to buy between 130 and 180 C-X aircraft, which will perform in both a strategic and a tactical (intratheater) role. However, because of the long lead time from concept to operational use, the first of these planes will not be in the force before the end of the decade. Finally, the navy will procure sixteen pre-positioned ships (AKX) between now and the end of the decade. If all of these programs are fully funded, the United States by the second half of the decade will be able to put a brigade-size force into the Persian Gulf area within a week, a division in about two weeks, and a 200,000-man force in about a month.

If the assets were applied strictly to Europe, the Pentagon by the end of the decade could almost meet its objective of placing five army divisions and nine tactical fighter squadrons into Europe within ten days after mobilization. It could not do both simultaneously, however. Nor could it be certain that the 12,000- to 200,000-man RDF or the enhanced forces in Europe could deal with the larger and better-equipped Soviet and Soviet-client forces they are likely to encounter. This is especially true for those forces destined for the Persian Gulf. DOD has not yet begun any program to increase the firepower of the light (unmechanized) forces assigned to the RDF.

There are no programs under way to improve our sealift forces or Merchant Marine, despite their present condition. The end of this decade is not likely to find a larger or newer defense reserve fleet or Merchant Marine.

When one compares overall U.S. modernization trends with those of the Soviet Union in the area of conventional forces, one is led to the same unfortunate conclusions that were found in regard to force structure: U.S. capabilities are declining relative to those of the Russians. Not only do they have a larger force structure now, they will have an even larger one in the future. Over the past seven years, the Soviets have introduced into their conventional forces four new classes of tactical fighters and a new class each of tanks, armored personnel carriers, sea control aircraft carriers, and amphibious ships. They have produced one new attack submarine per month (we produce one per year). Between now and 1985, the Soviets will begin construction of three nuclear-powered aircraft carriers,[11] will complete a class of 32,000-ton cruisers, will replace the forward echelons of their armored force in Central Europe with a more heavily armed and heavily armored tank, and will increase significantly their arsenal of surface-to-surface missiles. Given these trends, it is not surprising that, when they were asked by the Congress on May 29,

1980, about the adequacy of our current modernization programs, the joint chiefs of staff broke with tradition and responded unanimously that the programs were inadequate. General Edward C. Meyer, army chief of staff, put it quite simply: "I don't believe the current budget responds to Army's needs for the 1980s."

With current and projected force structure and modernization decidedly disadvantageous for the United States, the other two elements of capability—sustainability and readiness—become even more crucial.

4

SUSTAINABILITY

Stationing or moving forces into an area is one kind of challenge. Having sufficient replacements and stores to sustain the battle is quite another. This component of military capability is particularly important for a nation with global responsibilities if its forces are already suffering shortages of manpower and firepower before the battle even begins.

Unfortunately, sustainability is a major problem for every component of the undermanned and outgunned U.S. armed forces. If a full-scale Soviet attack were launched in the NATO area, our army ground forces could not sustain the battle for thirty days. Before that time, twenty-five days at the maximum, they would run critically short of tanks, armored personnel carriers, howitzers, artillery shells, tank rounds, mortar rounds, and bullets—in addition to a lack of qualified people. As mentioned above, the army is presently 22 percent short of the minimum number of tanks it needs and lacks almost 60,000 wheeled vehicles, which are required to move ammunition, fuel, food, and weapons. The war reserve stocks in Europe, depleted to support the Israelis in the 1973 Middle East war, have not yet been fully replaced. The active army, furthermore, does not have enough qualified people to man the augmented force in Europe. Within ninety days after mobilization for a NATO contingency, the army would have a 72 percent shortage of tank crews, a 48 percent shortage of infantrymen, and a 27 percent shortage of artillery troops. Finally, the deplorably understrength reserve, which is supposed to provide 90 percent of the replacements for battlefield casualties, has neither the numbers nor the training required for such a task. Of the 120,000 men in the Individual Ready Reserve (IRR) only 20,000 have the skills needed to replace casualties in armor and infantry units.[12] A spokesman for the National Guard Adjutants General Association, in noting that the guard and reserve units are woefully understaffed, underfunded, and unprepared for any military contingency, remarked in May 1980, "We couldn't mobilize enough firepower to stop Snow White and the Seven Dwarfs."

If the war should go beyond thirty days, the situation would get even worse for the army. This nation does not have the industrial base, the transportation capability, or the people to fight a prolonged war in Europe. It would take at least sixty days after mobilization for the complex of government-owned munitions plants to begin pumping out additional ammunition. Some items, like tanks and explosives, could not be produced more rapidly for two years because so many U.S. companies have gotten out of the defense business. The army reserve and National Guard are more than 180,000 below wartime strength; the IRR is short of people for the combat arms. Even the active army combat forces in the United States are now 15 percent below authorized levels.

Whatever men and materiel it could muster, the army would have an extremely difficult time getting "over there." Because of delays in moving the men and materiel to ports (caused primarily by the decrepit state of our rail lines), the difficulties of finding enough ship transportation, the threat posed by the 175 submarines that the Soviets could place in the North Atlantic, and the damage likely to occur to European ports (there are only seven European ports that can handle the cargo), it is extremely unlikely that sufficient supplies could arrive to sustain the war. Indeed, during the Nifty Nugget mobilization simulation exercise, conducted in the fall of 1978, some 350,000 tons of ammunition shipped to Europe by sea never arrived!

The other services are not in much better shape. The navy has severe shortages of torpedoes, missiles, and mines. Overall, according to Vice Admiral Staser Holcomb, director of navy program planning, it has about 50 percent of the ordnance it needs if it should become involved in a shooting war.[13] Most of its forward-deployed ships could not fight for more than a few days. Nor will things get much better in the near future. Present plans call for the navy to buy only 5,000 MK-46 torpedoes instead of the 9,000 needed,[14] only 2,000 Harpoon anti-ship vessels instead of the 3,000 needed; 2,000 MK-48 torpedoes instead of the required 4,000; and 900 mines instead of the 6,000 necessary. Because of these less-than-adequate programs, the navy's ordnance stockpiles will continue to be unable to sustain an acceptable level of combat.

The tactical aircraft of all the armed services and strategic airlift aircraft of the air force all suffer from a chronic lack of spare parts. The air force has only 15 percent of the war reserve spare parts that it needs. It is only through cannibalization of other planes that these forces maintain a minimum number of mission-capable planes. Moreover, the tactical fighters are also very low on air-to-air missiles. Officially the air force admits that it has only one-third of the required number of Sparrows and that the U.S. Air Force in Europe has enough missiles for only fourteen days of war. But the missile shortage in the air force is even worse than officially reported. Inventory objectives for air-to-air missiles assume that it would take only two missiles to destroy an enemy aircraft. In Vietnam, fighting against an enemy less sophisticated than the Soviets, it took eleven Sparrows or five Sidewinders to kill each enemy aircraft. Navy fighter aircraft are in even worse shape. The navy has on hand only two loadings of missiles for each of its fighter planes.[15] Finally, because of a lack of airlift capacity, the fighters that would be airlifted overseas could not even bring along all the armaments and spares that they do have in the inventory. It would take four to five weeks to deploy fully (from the United States to the Persian Gulf) the four tactical air wings assigned to the RDF.

To bring the 300 planes to their predeployment strength of spares and munitions would require one year. Similarly, about one-third of our giant C-5As, the only aircraft capable of carrying outsize cargo such as tanks, helicopters, and bulky radars, would have to be grounded within two weeks after mobilization because there are not enough spare parts for them in Europe.

The RDF also has severe problems in sustaining combat. To deploy a 100,000-man force into the Persian Gulf in response to a crisis situation, another 100,000 people would be required to provide support services. Most of these people are in the reserves, and the president is authorized to call up only 50,000. Assuming the support personnel could be found, the 100,000 forward-deployed people could not be supported without bases in the area. As Robert Komer, the undersecretary of defense for policy, has noted, "the viability of the RDF depends critically on our access to facilities in the area." Right now the closest U.S. base in the area is Diego Garcia, a tiny British-owned atoll on which the United States has developed an air and naval base. But Diego Garcia is 2,000 miles from the Persian Gulf, too far to provide logistic support adequate to the needs of the RDF. The 100,000-man RDF needs 1.2 million gallons per day of jet fuel, diesel oil, and gasoline and 25,000 gallons of drinking water daily to sustain itself. Even if the United States could get full access to the bases in the immediate vicinity of the fighting, and this seems unlikely, it would take a minor logistic miracle to support a force of over 100,000 people for any length of time. Finally, even if all these obstacles could be overcome, the RDF, like any other conventional force components,

will not have sufficient ordnance available in its planned inventory to accomplish its task. The Marine Corps will be permitted to buy only 2,000 TOW anti-tank missiles rather than 3,000. Thus, the corps can have only two instead of three TOW shooters in each company, a potentially catastrophic shortfall against the heavy armor likely to be encountered in the Persian Gulf area.

Those citizens concerned with military readiness often ignore our force structure, modernization, and sustainability or tacitly assume that they are within tolerable bounds. How much more concerned would they be if they knew the precarious condition of our military force in those areas of military capability?

5

Readiness

A force that is deficient in size and sustainability can often compensate for these inadequacies through a high state of readiness, that is, having the proper mix of people and equipment to perform effectively at the initiation of hostilities. Historians generally agree that at the outset of World War II, the Germans had fewer tanks and men than the allies but succeeded on the field because they were more combat-ready. Readiness is also more critical for the United States because we, unlike the Soviet Union, are a great distance from the likely areas of hostility.

Readiness, unlike force structure or modernization, is difficult to measure. By its very nature readiness can only be known once the real battle starts. Any system used to measure readiness in peacetime can only do so through the use of surrogates or substitutes. There have been instances where units rated or measured as non-combat-ready have performed well in combat and vice versa.

The Pentagon places combat units into five categories of readiness:

- C-1: Fully combat-ready.
- C-2: Substantially combat-ready, that is, the unit has only minor deficiencies.
- C-3: Marginally combat-ready, that is, the unit has major deficiencies but still can perform its assigned missions.
- C-4: Not combat-ready because the unit has so many deficiencies that it cannot perform its wartime functions.
- C-5: Not combat-ready because the unit is undergoing a planned period of overhaul or maintenance. For example, the carrier *Saratoga* is in the Philadelphia Naval Yard having its service life extended by fifteen years.

The difference between C-4 and C-5 is that a unit in the latter group is undergoing routine or planned maintenance, while the C-4 unit breaks down at a time when it is supposed to be in a state of good repair.

Since February 1980, DOD has used a unit reporting (UNITREP) system to place units into one of the five categories. This system establishes standard criteria in four areas that compare the unit's actual resources with those considered necessary to perform its wartime mission. The four areas that are evaluated are personnel, equipment and supplies on hand, equipment readiness, and training.

Personnel

In this area the unit commanding officer compares the people he has in the organization with the number and type of people that he is supposed to have in the event of a war. The evaluation is made on the following basis:

Category	*C-1*	*C-2*	*C-3*	*C-4*
A. Strength	$\geq 90\%$	$\geq 80\%$	$\geq 70\%$	$<70\%$
B. Critical skills	$\geq 85\%$	$\geq 75\%$	$\geq 65\%$	$<65\%$
C. Senior strength (E5-E9)	$\geq 85\%$	$\geq 75\%$	$\geq 65\%$	$<65\%$

The lower percentage in either category A or B establishes the actual C-rating for personnel. Thus, if the unit is between 80 and 90 percent in strength (the equivalent of a C-2 classification) but between 65 and 75 percent in critical skills (the equivalent of a C-3), the unit is classified as C-3 in personnel. Category C is optional and can be used by the commanding officer if he feels that the number of senior noncommissioned or senior petty officers is critical to the effective operation of his unit. Should he elect to use this category, then it determines the personnel readiness rating, if it is lower than A or B.

Equipment and Supplies on Hand

In this area, the commanding officer compares the amount of combat-essential equipment (planes, missile launchers, tanks), support equipment (fueling trucks, wheeled vehicles), and organic supplies (spare parts) that he actually possesses with the total that is required for the unit's wartime mission. This measurement does not evaluate the condition of the equipment. The comparison for arriving at the readiness rating is made using the following criteria:

Category	*C-1*	*C-2*	*C-3*	*C-4*
A. Service-selected, combat-essential equipment possessed ÷ wartime requirement				
Aircraft	$\geq 90\%$	$\geq 80\%$	$\geq 60\%$	$<60\%$
Other than aircraft	$\geq 90\%$	$\geq 80\%$	$\geq 65\%$	$<65\%$
B. Service-selected, end items support equipment and supplies possessed ÷ wartime requirement	$\geq 90\%$	$\geq 80\%$	$\geq 65\%$	$<65\%$

The calculation of the C-rating for the resource area is accomplished by determining the percentage of service-selected wartime materials possessed by the unit at the time of reporting and dividing it by the requirement in wartime. The lower percentage for either category A or B determines the rating. Thus, if a tactical aircraft squadron has between 80 and 90 percent of its required aircraft (C-2), but the squadron has less than 65 percent of its necessary spares (C-4), then the squadron is C-4, not combat-ready. It is important to note that the criterion for aircraft is less stringent in the C-3 and C-4 categories than for other major equipment. Thus, an army division with 64 percent of its required tanks would be C-4 or not combat-ready, while an air force fighter squadron with only 60 percent of its authorized complement would be C-3 or marginally combat-ready.

Equipment Readiness

In this area, the commanding officer evaluates the actual condition or the quality of the weapon systems and supporting elements in his unit. As in the other areas, he does this by comparing the actual condition of his equipment with that prescribed by the service for a wartime situation. Equipment readiness is determined in accordance with the following formula:

Category	*C-1*	*C-2*	*C-3*	*C-4*
A. Major equipment				
Aircraft	≥75%	≥60%	≥50%	<50%
Other units	≥90%	≥70%	≥60%	<60%
B. Mission-essential equipment	≥90%	≥70%	≥60%	<60%

The calculation for the C-rating in the area of equipment readiness is determined by applying the lower rating for either A or B. Thus, a nonaviation unit with 61 percent of its mission-essential equipment operationally ready but only 59 percent of its major equipment ready would receive an overall rating of C-4. For an aviation unit to receive a C-4 or not-combat-ready rating, more than half of its aircraft would have to be unable to fly.

Training

This is the most complex area of readiness to quantify. Under the UNITREP system the commanding officer is asked to evaluate his level of training by employing one of three criteria. He can estimate the time required for his present unit to achieve what his service defines as full training; or if he commands an aviation unit, he can compare the number of combat-ready air crews formed and available with the number required in wartime; or he can evaluate the percentage of full training that his unit has completed. Nonaviation units can use either the first or third option; air squadrons can choose any of the three options. The C-rating is determined as follows:

Category	*C-1*	*C-2*	*C-3*	*C-4*
A. Time in weeks to achieve full training	≤ 2	≤ 4	≤ 6	> 6
B. Or combat-ready crews	$\geq 85\%$	$\geq 70\%$	$\geq 55\%$	$< 55\%$
C. Or percentage of training complete	$\geq 85\%$	$\geq 70\%$	$\geq 55\%$	$< 55\%$

Under this system an aviation unit can be judged combat-ready (C-3) even if but 55 percent of its crews are combat-ready and if it will take more than six weeks to upgrade the remaining 45 percent of the crews. Likewise, any unit that has at least 55 percent of its training complete can be categorized as combat-ready even if the remaining training cannot be completed in six weeks or, if it is an aviation unit, if less than half of its crews are combat-ready.

The actual combat-readiness of a unit is determined by the lowest grade in the four categories just discussed. Thus, an army or marine division that is C-2 in personnel, C-3 in equipment on hand, C-1 in equipment readiness, and C-2 in training would receive an overall rating of C-3 or marginally combat-ready. Similarly, a ship that is C-1 in equipment on hand, equipment readiness, and training, but C-4 in personnel, because it has less than 65 percent of its critical skills, is C-4 or not-combat-ready. On the other hand, a unit could be combat-ready and still lack 30 percent of its people, 35 percent of its critical skills, 40 percent of its aircraft, 35 percent of its spares, and 45 percent of the necessary training!

Before looking at the actual readiness ratings of U.S. forces for mid-1980, five points must be kept in mind.

• First, the individual services determine, subject to DOD approval, what constitute the criteria against which the commanding officer judges his unit. These criteria are somewhat arbitrary and, therefore, are subject to varying degrees of interpretation. One could determine, for example, the number of missiles that a unit should possess by predicating that it needs to have a 95 percent confidence level of destroying a given number of targets. However, by dropping the confidence level to 90 percent, the number of required missiles can be reduced significantly.

• Second, readiness is an ephemeral indicator; it can and does change almost daily. For example, a twelve-plane attack squadron can be C-4 or not-combat-ready today if five of its planes are operational. If tomorrow only one of the seven grounded planes is repaired, the unit then becomes combat-ready.

• Third, each of the services measures different-size units. The army and Marine Corps measure readiness by divisions that can number up to 40,000 people; the navy measures ships and air squadrons that can have as few as 100 people; the air force measures readiness by squadrons that have several hundred people.

• Fourth, a unit is combat-ready or mission-capable if it is C-3 or above. This allows many units with barely half of their required assets to be judged mission-capable.

• Fifth, and perhaps most important, the UNITREP system, like any system that attempts to measure effectiveness through the use of surrogates, is subject to intentional or unintentional abuse. Such things as a "can do" spirit, pride of command, fear of being relieved, careerism, or just plain dread of passing bad news to the top can make it very difficult for officers in the field to inform headquarters that they cannot carry out their assigned mission. Apparently, this is what happened to the First Tactical Air Wing at Langley Air Force Base in June 1980. This unit, the most prestigious fighter unit in the air force and one of the first U.S.-based units slated to go to Europe or the Middle East in an emergency, was reporting that it was operationally ready. Nonetheless, a surprise inspection found that only 35 percent of its aircraft were actually mission-capable; that is, the unit was fifteen percentage points below the C-3 standard.[16]

Similarly, data can be manipulated to make units appear more combat-ready than they really are. For example, the number of companies in a division can be reduced; a non-combat-ready ship can be placed into a C-5 category, even though it was not supposed to be, and thus can be dropped from the readiness reporting system; the criteria for certifying an air crew as combat-ready can be altered; or an equipment objective can be reduced by such steps as changing the probability of a weapon's destroying a target.

Nonetheless, no matter how imperfect our measure of readiness may be, we must attempt such an assessment because if and when hostilities break out it will be too late to discover that we were not prepared. However, interviews with officers in the field indicate that, if anything, the reported figures are probably inflated.

The Department of Defense measures readiness by theater: Europe, Pacific, and the Continental United States (CONUS). The vast majority of the units in the Pacific and all of the units in Europe are considered forward deployed and are supposed to be ready to engage in combat at any time. Therefore, they are expected to be C-1 or C-2. Active units in the CONUS, which are targeted for either of the theaters, should also be operationally ready, but a high state of readiness is not as critical for them as for the forward-deployed units. Overall, DOD should have an absolute minimum of 70 percent of its units combat-ready at all times.

Table 1 displays the readiness of our major operational forces by theater as of mid-1980. As that table shows, about one-third of all the major combat units in DOD are not ready to fight as measured by DOD's own system. Some

TABLE 1

READINESS OF MAJOR OPERATIONAL UNITS BY THEATER, MID-1980
(percent)

Theater	Readiness Category C-1	C-2	C-3	C-4	C-5	Percentage of DOD Units Reporting
Europe	4	39	32	24	1	12
Pacific[a]	5	31	33	12	19	43
Continental United States[b]	5	32	26	17	20	45
Overall	6	32	30	15	17	100

NOTE: Major operational units include only divisions, ships, fighters, and attack and airlift squadrons.

[a] Includes all units assigned to the area of the commander-in-chief of the Pacific (CINCPAC).

[b] Includes navy ships assigned to the Atlantic fleet, all airlift squadrons, and one marine division–air wing combination.

SOURCES: Data for tables 1, 2, and 3 are derived from a great many sources. These include: John Fialka, "Air Force Having Trouble Keeping F-15 Jets Airborne," *Washington Star,* June 29, 1980; Richard Halloran, "Combat Readiness of F-15 Questioned," *New York Times,* December 11, 1979; Fred Hoffman, "Military Told to Overhaul Existing Gear," *Washington Post,* May 17, 1980; U.S. Congress, House of Representatives, Committee on Armed Services, *Review of Operation and Maintenance Budgets and Readiness,* 96th Congress, 1st session, December 18-19, 1979; Drew Middleton, "U.S. Air Readiness in Europe Declines," *New York Times,* July 13, 1980; Edward O'Brien, "Lack of Money, Men Is Limiting Readiness of U.S. Armed Forces," *St. Louis Globe-Democrat,* June 21-22, 1980; Clarence Robinson, "Military Stresses Funding Lack," *Aviation Week and Space Technology,* June 30, 1980; George Wilson, "Shortages of Parts Hamstring Warplanes," *Washington Post,* March 15, 1980; Richard Barnard, "Generals Warned of Helicopter Failures Prior to Iran Rescue Mission, but Did They Listen?" *Defense Week,* July 28, 1980; "The State of U.S. Military Readiness," *National Security Record,* July 1980; John Fialka, "Army's Supply, Maintenance Ability Hurt by Lack of Funds, General Says," *Washington Star,* August 27, 1980; and Raymond Coffey, "U.S. Military: Too Weak for War," *Chicago Tribune,* August 26-27, 1980.

17 percent of DOD's units are in planned overhaul while another 15 percent have a major deficiency in at least one area that is so serious that they are judged unable to carry out their combat mission. Moreover, 30 percent of all DOD units are C-3, only marginally combat-ready, that is, they have major deficiencies but can still go into combat. The frightening aspect of table 1 is that it demonstrates that less than 40 percent of all the divisions, planes, and ships in our armed forces are fully or substantially combat-ready—units in which the American people can have a reasonable amount of confidence that they can carry out their wartime mission. Moreover, only 6 percent of all the units in our entire armed forces have all the equipment, training, and personnel that they need. In 1979, the fully mission-capable rate (C-1 or C-2)

for the RH-53 D, Sea Stallion helicopters, which were used in the aborted rescue mission to Iran, was 17 percent![17]

When one considers differences among the theaters, the results are even more alarming. One quarter of our front line forces in Europe, the forces that must repel a Warsaw Pact attack, are not combat-ready, while another 32 percent of those forward-deployed units are in the marginal or risky category. Only 43 percent of the American units assigned to the European theater are really ready to go. One type of unit in Europe currently experiencing severe readiness problems is the air force tactical squadrons. Nearly half of the thirty air force squadrons in Europe, which are the largest single air component of NATO's deterrent forces, are not combat-ready. The U.S. Army in Europe also has problems. Its readiness is being undermined by its $1 billion backlog of needed maintenance and repairs.

Our readiness posture in the Pacific theater is even worse than in Europe. As table 1 indicates, approximately 31 percent of the units in that area are not combat-ready, while another 33 percent are rated as only marginally capable. That leaves only 36 percent of the entire force in the Pacific without some major problems.

This is a major change from even five years ago, when ships and squadrons deployed to the Pacific were never below C-1 or C-2. Today it is common. Moreover, the situation is getting worse instead of better. During the first third of 1980, the number of ships being deployed with a C-4 rating tripled!

As might be expected, our CONUS-based units have the highest percentage of combat units not ready, about 37 percent in the C-4 and C-5 category. It is important to keep in mind, however, that even some of these CONUS units, like the army's 82nd Airborne and 101st Air Mobile Divisions, the East Coast Marine Division, or our strategic airlift assets, which form the backbone of our Rapid Deployment Force, must be in as high a state of readiness as our forward-deployed units. Some of the CONUS units currently in a C-4 status include the 101st Air Mobile Division and the First Tactical Fighter Wing.

Table 2 breaks down current operational readiness by service. As that table indicates, the navy has the most units to report on since it must keep track of some 400 ships and seventy fighter and attack squadrons, while the army has the least since it only reports on sixteen divisions. Overall, except for the Marine Corps, none of the services is in very good shape. The navy and army are experiencing the most readiness problems. Only 31 percent of the navy's ships and combat aircraft are in good shape. Some 36 percent are not combat-ready while another 33 percent are combat-ready but have major problems. The situation in the navy has become so bad that it is routinely deploying C-4 ships and last spring actually had to tie up a ship because of personnel shortages. Only six of the army's sixteen divisions are without major difficulties. While the air force does not have as many prob-

TABLE 2

READINESS OF UNITS BY SERVICE, MID-1980
(percent)

Service	*Readiness Category*					*Percentage of DOD Units Reporting*
	C-1	C-2	C-3	C-4	C-5	
Army	0	37	25	38	0	3
Navy	3	28	33	15	21	76
Air force	18	39	18	16	9	18
Marine Corps	0	70	30	0	0	3
Total	6	32	30	15	17	100

SOURCES: Same as table 1.

lems as the army and navy, it still has more than 40 percent of its units with serious deficiencies. The Marine Corps is currently reporting that all of its divisions and air squadrons are combat-ready, but one must remember that the Marine Corps has been "C-2 for 200 years." One must therefore take such reports with a degree of skepticism.

Table 3 displays the readiness of our major combat units. As that table shows, significant portions of our ground divisions, tactical air squadrons, and aircraft carriers have severe readiness problems. Only one of our twelve aircraft carriers, the bulwark of our projection forces, lacks major deficiencies. Six are not fit for combat, and the remaining five only marginally capable of carrying out their assigned missions. It is no wonder that early this year the navy was forced to send a carrier rated C-4 to the critical Indian Ocean area, or that the carrier *Kennedy* became "ready" to deploy to the Mediterranean this summer only by "borrowing" fifty people with special skills from other ships. Similarly, about one-third of our nineteen ground divisions are not ready for combat, and less than half are free from major deficiencies in at least one

TABLE 3

READINESS OF MAJOR UNITS, MID-1980

	Readiness Category					*Total*
	C-1	C-2	C-3	C-4	C-5	
Ground divisions	0	9	4	6	0	19
Tactical air squadrons	15	47	56	24	19	161
Aircraft carriers	0	1	5	4	2	12

SOURCES: Same as table 1.

area. It is not surprising the the chief of staff of the army refers to many of his divisions as "hollow." Tactical aviation, which is in somewhat better shape than the carriers and ground divisions, still has only 62 of its 161 squadrons without major deficiencies and 43 units not combat-ready. Last year the mission-capable rate (C-1, C-2, or C-3) for the F-15, the air force's first-line fighter, was only 55 percent, while the mission-capable rate for the navy's new sophisticated F-14 fighter was only 53 percent.

The readiness problems of the army and the navy are primarily personnel-related. As our earlier study on military compensation showed, each of those services has shortages of noncommissioned officers in excess of 20,000 and thus is forced to man many skilled ratings at 65 to 75 percent. All army units in CONUS, except the 82nd Airborne, are under strength. The navy is also beginning to experience severe shortages among its pilots. A navy attack squadron should have twenty pilots; this year it has sixteen, and next year the number will drop to fifteen.

The navy's overall retention situation appears to be getting worse. Despite the severe unemployment problems besetting the nation, navy reenlistment rates for the first three quarters of FY 1980 were more than two percentage points below those of a comparable period in FY 1979, that is, down from 52.2 percent to 50.0 percent.

Army reenlistment rates for FY 1980 are slightly higher than a year ago. However, the army has severe quality problems with those that it does recruit and retain. Only 38 percent of the people joining the army in the first three quarters of FY 1980 had a high school diploma, and 45 percent of those who joined in FY 1979 were classified in the lowest intelligence category, that is, category IV. The result of the army's own 1979 skill qualifications tests were horrendous. Overall, about half of the soldiers taking various job skill tests failed to pass. In ten skills, the failure rate was over 90 percent. For example, 98 percent of tank repairmen failed, as did 91 percent of the aviation maintenance personnel, 86 percent of artillery crewmen, 83 percent of transportation personnel, 69 percent of the communications people, 51 percent of the military intelligence personnel, 49 percent of the combat engineers, and 43 percent of the field artillerymen. Only three bassoon players, eleven special agents, and sixteen missile mechanics were able to constitute skill groups passing their tests by 100 percent.[18]

A comprehensive twelve-volume army study on the volunteer military, which was completed in February 1979, concluded that the intelligence level of the crew members manning tanks and air defense systems is too low to operate them at army standards for combat performance.[19] The study also found that over 20 percent of the tank gunners in Europe did not know how to aim the gun and that less than 20 percent of the tank repairmen could diagnose and repair broken tanks. The skill level has become so low that the army is being forced to rewrite training manuals at the grade school level. To put it bluntly, the army does not have enough quality people to be a completely combat-ready force.

Both the army and the navy are experiencing training problems as well. The shortage of fuel and the lack of training weapons have made it difficult for the army and navy to give their people adequate training. Soldiers in Europe are lucky if they get to fire one anti-tank missile per year. Similarly, the fuel that powers the Mediterranean fleet is 20 percent less than was allotted to the fleet last year and 30 percent less than it was five years ago. In FY 1980, overall steaming days at sea for the entire navy are 17 percent below minimum levels, while tactical aircraft operations will fall 15 percent short.[20]

While the air force has some personnel problems, including a shortage of 3,000 pilots, they are not yet as severe as those of the army and navy. The trends are disturbing, however. During the first three quarters of FY 1980, the air force reenlistment was down to 60.9 percent. If this trend continues into the last quarter, it will mark the third year in a row that the rate has dropped and will place air force reenlistment rates below those of the army. The majority of air force readiness problems are presently caused by lack of spare parts and inadequate funding. The dollar value of the current air force spare parts shortage is $2 billion. The average number of flight hours in an air force flight crew has declined by 56 percent in the last decade, as the amount of fuel consumed annually by the service has dropped by 55 million barrels since 1973. Air force combat pilots should fly at least sixteen hours per month; they fly only ten. Only 1 in 100 air pilots in Europe can participate in training exercises run by the Tactical Air Command. The French actually send more pilots to these exercises than the American forces in Europe. Air force pilots should fire one missile a year in training. Today, some are lucky to fire one in their entire careers. With situations like this, is it any wonder that the air force is currently retaining only 31 percent of the pilots with six to eleven years in the service?[21]

Although the Marine Corps is reporting "100" percent of its units combat-ready, it too has some personnel shortages. It currently lacks some 3,000 technicians.[22] Moreover, its reenlistment rate for FY 1979 was the lowest of the four armed services, 34.9 percent compared with 57.4 percent for the other three services. The corps is also in need of additional equipment, like anti-tank weapons. Finally, it is important to note that even a C-2 Marine Corps cannot be very effective without adequate lift.

It is clear that no matter how one evaluates readiness—by theater, service, weapon system, or DOD-wide—or how imperfect the measuring system, the current readiness of our military units is unsatisfactory. It is no wonder that the current secretary of defense is not satisfied with the combat readiness of our forces. In the *Consolidated Guidance* for the FY 1982 defense budget and the FY 1982–1986 defense program, Secretary of Defense Brown belatedly gave strong orders to improve readiness.

6

Summary

Overall, the current capability of our armed forces to carry out their assigned mission is inadequate. As this study shows, our force structure is too small, our modernization programs are insufficient, and our active-duty operational forces have severe readiness and sustainability problems. If our armed forces have to go into combat in the immediate future, there must be considerable doubt about their ability to perform adequately.

How did this situation develop, and what is to be done about it?

To understand how we got into this predicament, it is necessary to go back fifteen years, to the height of the American involvement in Southeast Asia. From 1965 through 1968, as American involvement increased, the costs of conducting the war rose rapidly. (By 1968, they came to about $67 billion in today's dollars.) In order to hold down the costs of the overall defense budget during that period, capital investment was cut back drastically. For example, the number of ships that the navy bought during the war was reduced by 80 percent from its pre-1965 levels. From 1969 through 1975, the Congress and the American people demanded that defense expenditures be reduced in real terms. During this same period, personnel and operating costs rose rapidly. Increased personnel costs resulted from three factors: the adoption of comparability for military and civilian employees of the federal government, the rapid growth of the retired population, and the creation of the all-volunteer force. Between FY 1968 and FY 1976, the share of personnel costs rose from 42 to 55 percent of the budget. Higher operating costs came about primarily because of the drastic increase in the price of the 200 million barrels of oil used annually by the Pentagon. Between FY 1968 and 1976 operating costs per unit more than doubled. Consequently, DOD had to cut back its force structure, reduce training, slow down force modernization, and accept shortfalls in spare parts and war reserves in order to accommodate decreasing budgets.

Since 1975, the slight real rise in the defense budget has gone primarily toward modernization of the force structure. Funds for personnel, maintenance, training, and sustainability have remained constant or have been reduced. This situation is vividly demonstrated by the data contained in table 4. As that table shows, over the past five years, personnel expenditures have grown by just over 3 percent in real terms, that is, less than 1 percent per year. Similarly, operating costs have grown by just over $2 billion or 9.3 percent, an increase of less than 2 percent per year. Expenditures for

TABLE 4

COMPONENTS OF THE DEFENSE BUDGET, FISCAL 1976 AND 1981
(in billions of constant fiscal 1981 dollars)

	Fiscal Year		Change	
Category	1976	1981	Dollars	Percentage
Personnel	68.6	70.7	2.1	3.1
Operating	22.5	24.6	2.1	9.3
Investment	33.7	47.4	13.7	40.6
Total	124.8	142.7	17.9	14.3

SOURCE: Harold Brown, *Annual Defense Department Report, FY 1981* (Washington, D.C., January 29, 1980); and Assistant Secretary of Defense (Comptroller), *National Defense Budget Estimates for FY 1981.*

investment, on the other hand, have increased over 40 percent, a real increase of 8 percent per year. This is thirteen times more rapid than the increase in personnel expenditures and four times more than the increase in operating costs.

Since 1965 the Soviet Union has increased dramatically the resources that it has devoted to defense. Since 1968 they have spent a quarter of a trillion dollars more on defense than has the United States. In the decade of the 1970s alone, the Soviets spent over $100 billion more on investment than we did. In FY 1980 they will spend $50 billion more on defense than the United States. With this difference in emphasis, there is no reason to wonder why the current and future Soviet force structure and sustainability far exceed ours. While we do not have precise readiness figures on all the components of the Russian forces, we do know that all Soviet ground divisions outside the U.S.S.R. are category A, that is, they have at least 75 percent of their allotted manpower.[23] Moreover, we know also that the units that they sent into Afghanistan were not part of their front-line troops, which are the ones kept in the highest state of readiness. Yet these invading forces have executed their military tasks quite well.

Nor will the situation change very much over the next few years. Between now and 1985 U.S. investment expenditures are projected to rise by $20 billion in real terms, from $50 billion to $70 billion. During that same period spending on investment by the Soviet Union will increase by $30 billion in real terms, from $90 billion to $120 billion.[24] Thus, over the next five years, the Soviets will spend on investment, in FY 1981 dollars, $520 billion while the United States will spend $280 billion, a difference of $240 billion or 46 percent.

The answer to the current crisis in military preparedness must begin with money. While this cannot be the final answer, it is a necessary precondition to correcting our current dangerous situation. The solution is not to cut back procurement and put the funds into training and maintenance. As the foregoing analysis shows, the new equipment is badly needed. To purchase combat readiness at the expense of future force structure would only solve one problem by creating another. What is needed is a substantial increase in the size of the program proposed last January by the administration. Rather than the current level of just over 5 percent of the gross national product (GNP), we should be willing to spend at least 6 percent of our GNP on defense. While this is not an inconsiderable sum, it is less than half of what the Soviets spend and it is less than the percentage we gave to defense throughout most of the post–World War II period. The key issue is not whether we can afford to spend the additional money, but whether we can afford not to spend it. There is no doubt that the 1980s are going to be dominated by considerations of foreign and military policy and that an ill-prepared, ill-equipped, undermanned U.S. military will make it difficult for the United States to deter war, to exercise leadership, and to compete successfully.

About one-third of the additional funds that will come from increasing the portion of GNP spent on defense should be spent on increasing pay, benefits, and special bonuses in order to ensure comparability, competitiveness, and equity for our military people. The initiatives undertaken by Congress this year (for example, the 11.7 percent pay raise and the Nunn-Warner bill) will help somewhat. But much more must be done. Retaining experienced noncommissioned officers and middle-grade officers with special skills is one of the keys to enhancing readiness.

Another one-third of the funds should go toward reducing the maintenance backlog, increasing training time, buying more training weapons, and purchasing additional spare parts. The most sophisticated weapon systems in the world cannot perform effectively without proper maintenance, sufficient spares, or trained people. Nor will the trained people stay around very long if they cannot operate their equipment.

The final one-third should go toward increasing the sustainability of our forces. We ought to have sufficient prepositioned materiel and war reserves to fight for at least ninety days in Europe. The tactical air units and ships outside Europe should have enough equipment on hand to conduct high intensity combat for at least one month. To provide our forces with any less is not only militarily but morally indefensible.

In addition to increasing our own defense budget, we must insist that our allies do more to provide for the common defense. We simply do not have the economic wherewithal, absolutely or relatively, to continue to bear a disproportionate share of the defense burden. We now spend twice as much per capita on defense as do our allies. If our major allies each put 5 percent of their GNP into defense, $60 billion additional would be available for defense in FY 1981. The NATO countries ought to provide more of the ground and air power on the Central Front and on the flanks of Europe, while the Japanese have to take more responsibility for sea control in the Western Pacific. For its part, the United States ought to allocate more of its limited forces toward the Middle East and Persian Gulf where the interests of the allies are as involved as ours. Coupled with the strategic umbrella provided by the United States, this ought to result in a more equitable distribution of the defense burden plus an enhanced defense posture.

Finally, we should look into ways to multiply our own military capabilities. For example, we need to reestablish a worldwide system of forward military bases for air and sea forces. We particularly need bases in the Indian Ocean, Australia, and Thailand. Homeporting of another carrier overseas, perhaps in Portugal, would also help make our present force structure more capable.

If for some reason the additional funds are not forthcoming and we are unable to persuade our allies to do more, then we must pursue a balanced policy. Balance is the key to success on the battlefield as well as the athletic field. Current readiness cannot be sacrificed to future capability. Given the current state of international affairs, the future may be now. Sufficient spare parts, proper maintenance, adequate training, and qualified personnel must be procured simultaneously with new weapons systems. Better a smaller force of F-15s, F-14s, or XM-1 tanks than large numbers of these systems rated not combat-ready. Better thirteen combat-ready divisions than sixteen hollow ones; or twenty-two C-2 air wings than twenty-six rated C-3 and C-4; or 350 general purpose ships than 450 operated by ill-trained, inexperienced people.

We hope the situation will not come to that. In contrast to the wishful thinking of some during the past few years, perhaps there is now belated recognition of the need for clear thinking about the costs of a sound defense and military readiness. We are confident that, as in the past, the American people, if told the truth, will meet the challenge. The military readiness of American armed forces cannot be left in its present state.

Notes to Text

[1] International Institute for Strategic Studies, *The Military Balance, 1979–80,* pp. 86–88; and *FY 1981 Defense Report,* pp. 71, 87, 89, 123–25.

[2] A Soviet division has about three-quarters of the number of soldiers of an American division but has the same amount of firepower; figures on the NATO balance can be obtained from U.S. Department of State, *Some Soviet and Warsaw Pact Military Facts and Figures,* January 1979; and International Institute for Strategic Studies, *The Military Balance, 1979–80,* pp. 3–30, 108–13.

[3] *FY 1981 Defense Report,* p. 155.

[4] This figure is based upon maintaining a force of two battle groups continuously deployed in each ocean.

[5] See, for example, the statement of Vice Admiral Staser Holcomb, director of navy program planning, in Ann Hughey, "The Age of Aircraft Carrier Diplomacy," *Forbes,* July 21, 1980, p. 57.

[6] "Reasonable assurance" is a term used by the Joint Chiefs of Staff in the Joint Strategic Planning System to describe the number of weapon systems needed to have a realistic chance of achieving a particular objective. These figures are contained in L. Edgar Prina, "Joint Chiefs Call for Major Boost in Combat Ships," *San Diego Union,* July 19, 1980, p. 1.

[7] These figures are derived from U.S. Congress, Congressional Budget Office, *U.S. Airlift Forces,* 1979, p. 55. See also Deborah Shapley, "Rapid Deployment Plan Raises Basic Procurement, Transport Questions," *Defense Policy Report,* June 20, 1980, p. 12; Thomas Toch, "Rapid Deployment: A Questionable Trump," *Baltimore Sun,* July 6, 1980, p. K1; and John Fialka, "Pentagon Outlines U.S. Response in Gulf," *Washington Star,* February 3, 1980, p. 8.

[8] See, for example, the statements of Leslie Gelb, senior fellow at the Carnegie Endowment for International Peace and former director of political-military affairs at the State Department, and Martin Binkin, a senior fellow at the Brookings Institution. Both are quoted by Toch, "Rapid Deployment."

[9] According to Dr. William Perry, undersecretary of defense for research and engineering, the Soviet Union will build 2,000 T-80 tanks per year. "Maximum Planned U.S. Tank Production/Half of Soviets," *Defense Daily,* July 16, 1980, p. 74.

[10] *Statement of Admiral Thomas B. Hayward on the FY 1981 Military Posture and the FY 1981 Budget,* January 31, 1980, pp. 10–13.

[11] "Russians Said to Build First Aircraft Carrier with Nuclear Power," *New York Times,* August 14, 1980; and Michael MccGwire, "The Rationale for the Development of Soviet Seapower," *United States Naval Institute Proceedings,* May 1980, pp. 155–83.

[12] Data for this section came primarily from the *Report on the Nifty Nugget Exercise,* summarized in a four-part article by John Fialka, "The War Game: A Mobilization Snafu," *Washington Star,* November 2–4, 1979.

[13] Quoted in Ann Hughey, "Age of Aircraft Carrier Diplomacy," p. 59.

[14] Based upon a 99 percent confidence level that navy ships will have the firepower to destroy a given number of assigned targets. Until mid-1980 this was how the Department of Defense arrived at its procurement objectives. For a fuller discussion of this point, see Richard Barnard, "DOD Solves Its Ordnance Shortage by Fiddling the Numbers," *Defense Week,* June 16, 1980, pp. 1, 9.

[15] Richard Barnard, "High Cost and Poor Management Deplete Missile Inventories," *Defense Week,* July 2, 1980, p. 8; and George Wilson, "Missiles Too Costly for Practice by Pilots," *Washington Post,* June 21, 1980, p. 2.

[16] John Fialka, "Air Force Having Trouble Keeping F-15 Jets Airborne," *Washington Star,* June 29, 1980, p. 1.

[17] Richard Barnard, "Generals Warned of Helicopter Failures Prior to Iran Rescue Mission, but Did They Listen?" *Defense Week,* July 28, 1980, p. 6.

[18] George Wilson, "Pentagon and Senate Allies Wrangling," *Washington Post,* June 26, 1980, p. 8.

[19] The study is called the "Army Training Study." Its contents are summarized in John Fialka, "GI's Proficiency at Low Level, New Study Says," *Washington Star,* February 3, 1980, p. 1.

[20] *FY 1981 Defense Report,* pp. 187–88.

[21] "State of U.S. Military Readiness," *National Security Record,* July 1980, pp. 1–3; and Edward O'Brien, "Are We Ready?" *St. Louis Globe Democrat,* June 21–27, 1980.

[22] Melvin Laird, *People, Not Hardware: The Highest Defense Priority,* American Enterprise Institute Special Analysis, No. 80–1. Washington, D.C., 1980, p. 6.

[23] Department of State, *Some Soviet and Warsaw Pact Military Figures,* p. 5.

[24] *FY 1981 Defense Report,* p. 35.

Bibliography

Abrams, Arnold. "Money and Manpower the Main Issues." *Los Angeles Times,* June 25, 1980.

Bacon, Kenneth. "A Shortage of Skills Is Hurting the Armed Forces as Civilian Pay Lures Away Personnel." *Wall Street Journal,* March 25, 1980.

Barnard, Richard. "DOD Solves Its Ordnance Shortage by Fiddling the Numbers." *Defense Week,* June 16, 1980.

———. "The RDF: What's Been Done, What Should Have Been Done." *Defense Week,* June 30, 1980.

Brown, Harold. *Annual Defense Department Report, FY 1981.* Washington, D.C., January 29, 1980.

Burt, Richard. "Pentagon Spending for Year Cut So It Can Rise Later." *New York Times,* April 17, 1980.

———. "U.S. Aides Say Soviets May Look to Persian Gulf for Oil in 1980's." *New York Times,* April 15, 1980.

Congressional Budget Office. *Improving the Readiness of the Army Reserve and National Guard: A Framework for Debate.* Washington, D.C., February 1978.

———. *The Marine Corps in the 1980's: Prestocking Proposals, the Rapid Deployment Force and Other Issues.* Washington, D.C., May 1980.

———. *Real Growth and Decline in Defense Operating Costs: Fiscal Year 1978* (Staff Working Paper). Washington, D.C., June 1977.

———. *Strengthening NATO: Pomcus and Other Approaches.* Washington, D.C., February 1979.

———. *U.S. Air and Ground Conventional Forces for NATO: Mobility and Logistics Issues.* Washington, D.C., March 1978.

———. *U.S. Airlift Forces: Enhancement Alternatives for NATO and Non-NATO Contingencies.* Washington, D.C., April 1979.

Cooley, John. "Indian Ocean Crises Stretch Navy Uncomfortably Thin." *Christian Science Monitor,* March 31, 1980.

———. "U.S. Reaction Time to Persian Gulf Is Limited by Its Air, Sealift Capacity." *Christian Science Monitor,* February 4, 1980.

Corddry, Charles. "Lack of Skills Cripples Navy." *Baltimore Sun,* April 27, 1980.

———. "U.S. Demands an End to 'Lag' by NATO Allies." *Baltimore Sun,* April 24, 1980.

Evans, David. "Enlisting Trouble." *Washington Post,* June 26, 1980.

Fallows, James. "Muscle-Bound Superpower." *The Atlantic,* October 2, 1979.

Fialka, John. "Air Force Having Trouble Keeping F-15 Jets Airborne." *Washington Star,* June 29, 1980.

———. "Civilians Flunk, Too, in 'Nugget' Exercise." *Washington Star*, November 4, 1979.

———. "Ill-Equipped, Undermanned U.S. Army Is Decimated in Nifty-Nugget Exercise." *Washington Star,* November 2, 1979.

———. "Pentagon Logistics Bridge Fails." *Washington Star,* November 3, 1979.

———. "Rapid Deployment Meaningless without Fuel, Water." *Washington Star,* April 9, 1980.

———. "25% of Recruits in Low-Intelligence Category." *Washington Star,* March 11, 1980.

General Accounting Office. *Attrition in the Military—An Issue Needing Management Attention.* Washington, D.C.: February 20, 1980.

———. *Implications of Highly Sophisticated Weapon Systems on Military Capabilities.* Washington, D.C.: June 30, 1980.

Getler, Michael. "Allied Defense Costs: An Unequal Sharing of the Burden." *Washington Post,* May 20, 1980.

———. "F-18s Cost, Technical Troubles Spur Probe Demand." *Washington Post,* June 13, 1980.

———. "Gulf Buildup Is Stretching U.S. Forces around the World." *Washington Post,* July 19, 1980.

Halloran, Richard. "Armed Forces Chiefs Say Personnel Losses Weaken U.S. Defense." *New York Times,* March 22, 1980.

———. "Combat Readiness of F-15 Questioned." *New York Times*, December 11, 1979.

Hoffman, Fred. "Military Told to Overhaul Existing Gear." *Washington Post,* May 17, 1980.

International Institute for Strategic Studies. *The Military Balance 1979-80.* London, England: 1979.

Kelly, Orr. "A Trillion for Arms Just to Catch Up." *U.S. News and World Report,* July 21, 1980.

Korb, Lawrence. "The FY 1981-85 Defense Program: Issues and Trends." *AEI Foreign Policy and Defense Review* 2 (August 1980).

Laird, Melvin. *People, Not Hardware, the Highest Defense Priority.* American Enterprise Institute Special Analysis, No. 80-1. Washington, D.C.: American Enterprise Institute, 1980.

Middleton, Drew. "Navy Sees Limit on Ability in Atlantic War." *New York Times,* February 20, 1980.

———. "U.S. Air Readiness in Europe Declines." *New York Times,* July 13, 1980.

———. "U.S Military Can Match Soviet, Officials Say, but Not before 1990." *New York Times,* January 7, 1980.

O'Brien, Edward. "American Forces Face Stiff Test in Long War, Military Leaders Say." *St. Louis Globe-Democrat,* June 24, 1980.

———. "Base Commanders War: The Fight for Enough Money." *St. Louis Globe-Democrat,* June 25, 1980.

———. "80's Will Be Critical Decade for West, General Says." *St. Louis Globe-Democrat,* June 27, 1980.

———. "Lack of Money, Men Is Limiting Readiness of U.S. Armed Forces." *St. Louis Globe-Democrat,* June 21-22, 1980.

———. " 'Very Real' Problem in Europe, Report Says." *St. Louis Globe-Democrat,* June 23, 1980.

"The Price of Power." *Time,* October 29, 1979.

Record, Jeffrey. "Why Plan Rapid Deployment of the Wrong Kind of Force?" *Washington Star,* February 3, 1980.

Robinson, Clarence. "Military Stresses Funding Lack." *Aviation Week and Space Technology,* June 30, 1980.

Stevenson, John. "Navy Being Spread Thin to Handle Mideast Duty." *Norfolk Virginian-Pilot,* January 27, 1980.

United Press International. "U.S. Lacks Ships for War, Admiral Says." *Atlanta Journal,* March 5, 1980.

U.S. Congress, House of Representatives, Committee on Armed Services. *NATO Standardization, Interoperability, and Readiness.* 95th Congress, 2d session, No. 96-29, 1978.

———. *Review of Operation and Maintenance Budgets and Readiness.* 96th Congress, 1st session, December 18-19, 1979.

Wilson, George. "Joint Chiefs of Staff Break with Carter on Budget Planning for Defense Needs." *Washington Post,* May 30, 1980.

———. "Missiles Too Costly for Practice by Pilots." *Washington Post,* June 21, 1980.

———. "Pentagon and Senate Allies Wrangling." *Washington Post,* June 26, 1980.

———. "Shortages of Parts Hamstring Warplanes." *Washington Post,* March 15, 1980.

Woolsey, R. James. "Studying Defense to Death." *Washington Post*, July 3, 1980.

Yarmolinsky, Adam. "The Military Balance: About Equal." *New York Times,* February 4, 1980.

Yost, Robert. "Inviting War by On-and-Off Defense." *Business Week,* July 14, 1980.

The AEI Public Policy Project on National Defense

The American Enterprise Institute established the Public Policy Project on National Defense in June 1976 to examine the many controversial issues involved in the defense of the United States. The project commissions research into concepts of strategy, manpower and force structure, the economics and administration of defense, and other major aspects of defense policy. It also sponsors television debates and round table discussions, as well as informal seminars and conferences, on current defense issues.

The project is chaired by Melvin R. Laird, a former congressman, secretary of defense, and domestic counsellor to the President, and now a senior counsellor for *Reader's Digest.*

An advisory council composed of distinguished proponents of many different viewpoints participates in project programs, which are designed to encourage the competition of ideas on defense policy.